A STORY OF THE SOLAR ECLIPSE

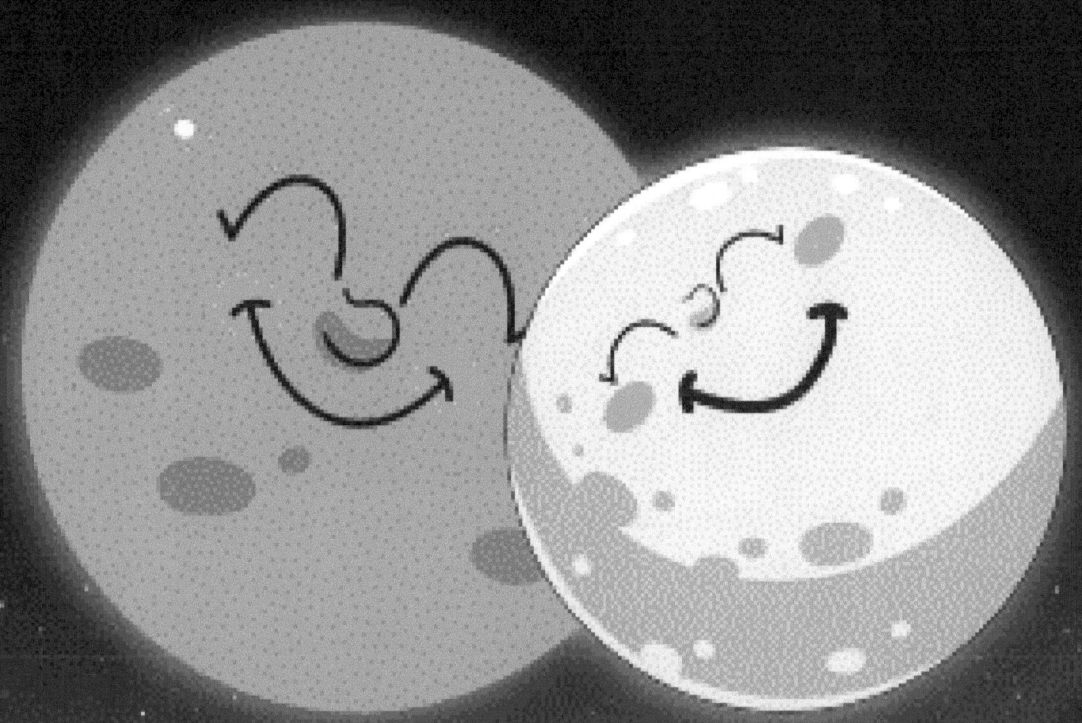

AN ASTRONOMY SOLAR ECLIPSE STORY

CHILSTORY

COPYRIGHT © 2024

ALL RIGHT RESEVED NO PART OF THIS PUBLICATION MAY BE REPRDUCED DISTRIBUTED PR TRANSMITTED IN ANY FROM OR BY ANY MEANS INCLUDING PHOTOCOPYING RECORDING OR OTHER ELECTRONIC OR MECHANICAL METHODS. WITHOUTTHE PRIOR WRITTEN PERMISSION OF THE PUBLISHER EXCEPT IN THE CASE OF CRITICAL REVIEWS AND CERTAIN OTHER MON COMMERCIAL ISERS PERMITTED BY COPYRIGHT LAW

INTRODUCTION

Welcome to "The Day the Sun Played Hide and Seek: A Solar Eclipse Adventure"! Join Timmy the brave rabbit, Lily the clever squirrel, and Benny the adventurous bear on a magical journey through the wonders of nature. In this enchanting tale, the trio discovers the mystery of a solar eclipse and learns about the importance of protecting their eyes. With the guidance of Wise Old Owl, they craft homemade eclipse viewers and embark on an unforgettable adventure filled with laughter, friendship, and wonder. Get ready to witness the magic of the celestial sky with Timmy, Lily, and Benny as they share their unforgettable experience with families and friends, leaving readers inspired to explore the world around them.

Welcome to the world of fantasy and excitement

In the heart of a lush green forest, nestled among towering trees and babbling brooks, there lived three adventurous friends: Maya the curious fox, Oliver the imaginative rabbit, and Sophie the wise owl.

One sunny morning, as they played hide-and-seek among the ferns, Maya spotted something peculiar in the sky. "Look!" she exclaimed, pointing to the sun. "It's disappearing!"

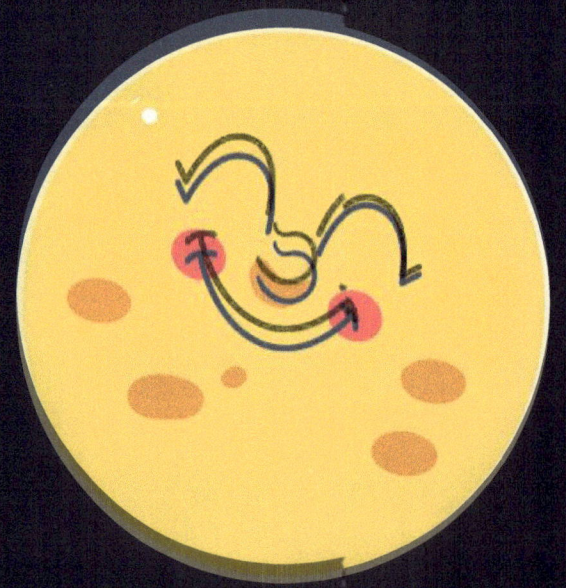

Oliver and Sophie rushed to her side, their eyes wide with wonder. They watched in amazement as the sun gradually vanished behind a dark shadow.

"What's happening?" Oliver asked, his voice filled with excitement tinged with a hint of uncertainty.

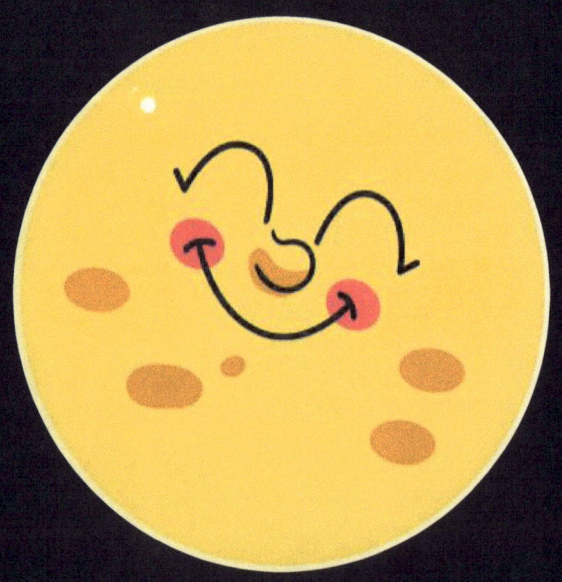

Sophie, with her vast knowledge of the forest's secrets, recognized the phenomenon. "It's a solar eclipse," she explained gently, her voice calm and reassuring.

Maya's ears perked up with curiosity. "But why does it happen?" she asked, eager to understand.

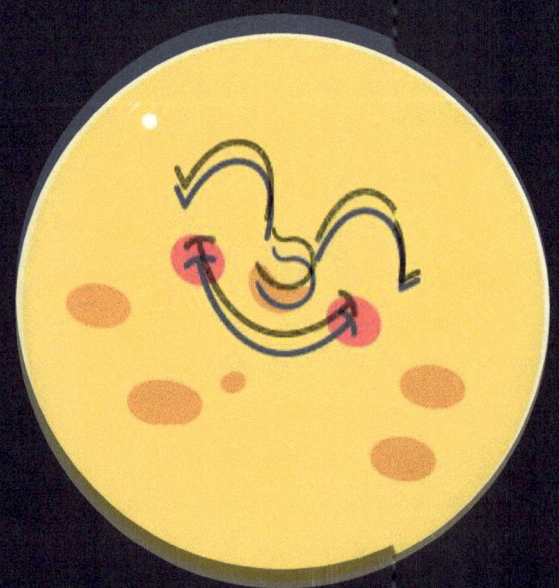

Sophie smiled patiently, ready to share her wisdom. "A solar eclipse happens when the moon passes between the sun and the Earth, casting a shadow on our planet," she explained. "It's a rare and magical event that reminds us of the wonders of the universe."

As they gazed up at the sky, Sophie reminded her friends of an important rule: "During a solar eclipse, it's crucial to protect your eyes. Looking directly at the sun can harm them."

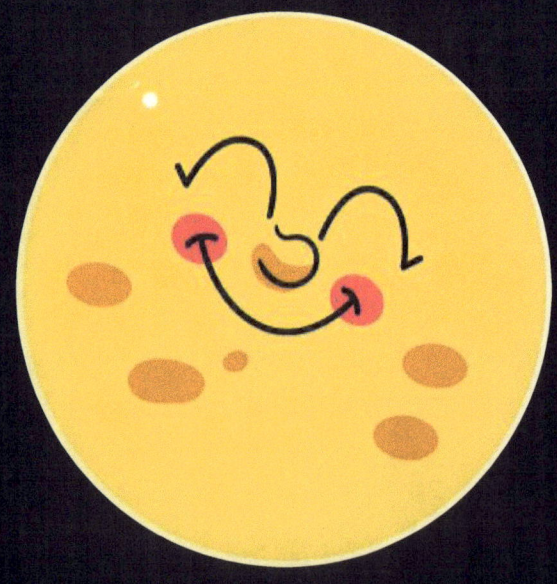

Maya nodded, her eyes wide with understanding. "How can we watch it safely?" she asked, eager to witness the spectacle without danger.

Oliver, always full of creative ideas, suggested they make their own eclipse viewers using cardboard boxes and aluminum foil. "We can make tiny pinholes to project the image of the eclipse onto the ground," he exclaimed.

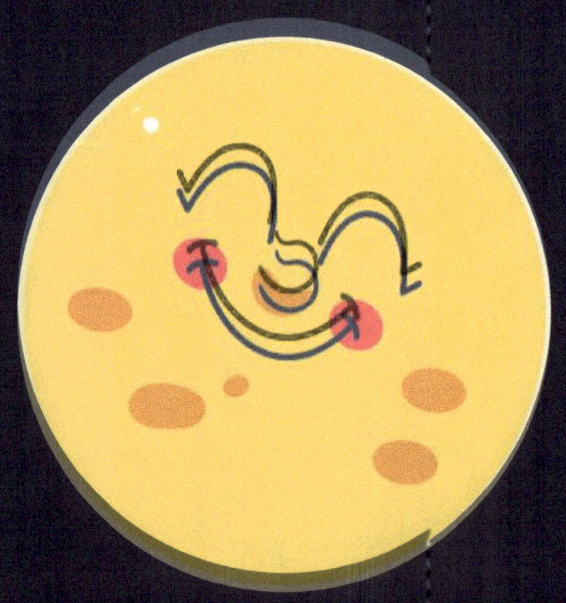

In the heart of a lush green forest, nestled among towering trees and babbling brooks, there lived three adventurous friends: Maya the curious fox, Oliver the imaginative rabbit, and Sophie the wise owl.

Excitedly, they gathered materials and crafted their viewers, following Oliver's ingenious design. With their homemade devices in hand, they ventured into a clearing to witness the eclipse safely.

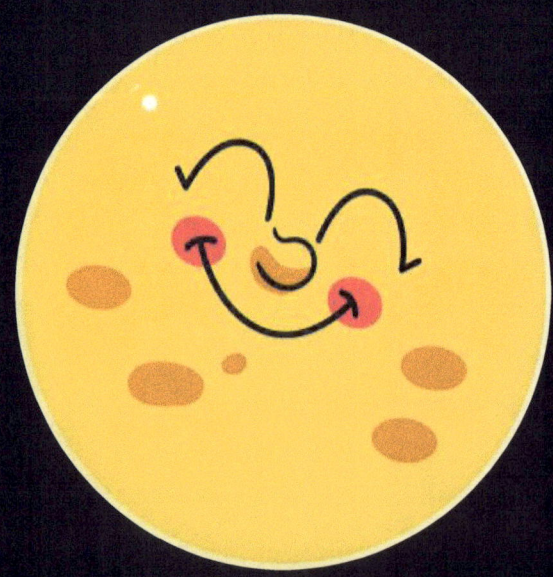

As they watched the moon slowly cover the sun, casting shadows on the forest floor, Maya, Oliver, and Sophie felt a sense of awe and wonder wash over them. They marveled at the beauty of the natural phenomenon, grateful for the opportunity to witness it together.

As the eclipse reached its peak and the sun began to emerge from behind the moon's shadow, they bid farewell to the magical sight, knowing that they had shared something truly special.

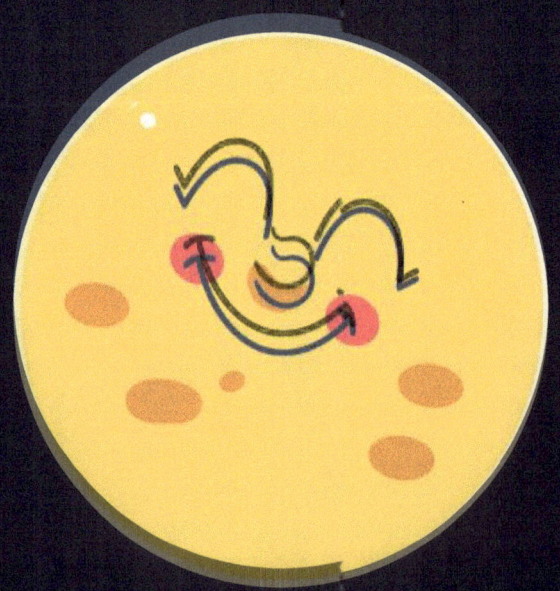

With hearts full of wonder and minds filled with newfound knowledge, Maya, Oliver, and Sophie returned to their forest home, eager to share their eclipse adventure with the other creatures of the woods.

The following morning, Maya, Oliver, and Sophie woke up to find the forest buzzing with excitement. Animals of all shapes and sizes gathered around, eager to hear about their eclipse adventure.

With enthusiasm, Maya recounted the tale, explaining the magic of the solar eclipse and the importance of protecting their eyes. Oliver added his creative flair, describing how they made their own viewers to safely watch the event.

Sophie, with her calm demeanor, emphasized the wonder of the universe and the value of learning from nature's mysteries.

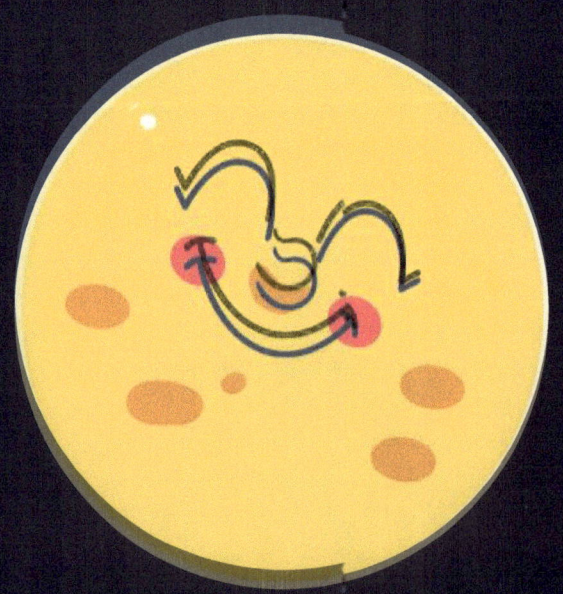

The animals listened intently, absorbing every detail of the story. Some nodded in understanding, while others gasped in amazement.

Inspired by Maya, Oliver, and Sophie's adventure, the animals began to ask questions and share their own observations of the eclipse.

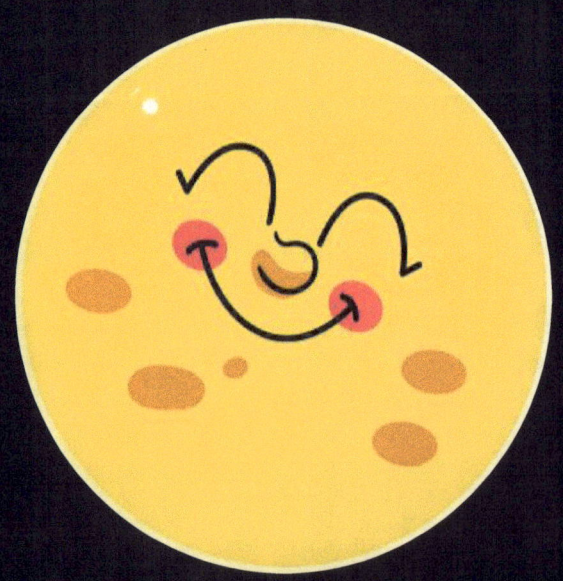

Together, they formed a circle of learning, exchanging knowledge and experiences, united by their curiosity and love for the natural world.

As the day drew to a close, Maya, Oliver, and Sophie felt a sense of pride knowing that they had sparked a newfound interest in the wonders of the universe among their friends.

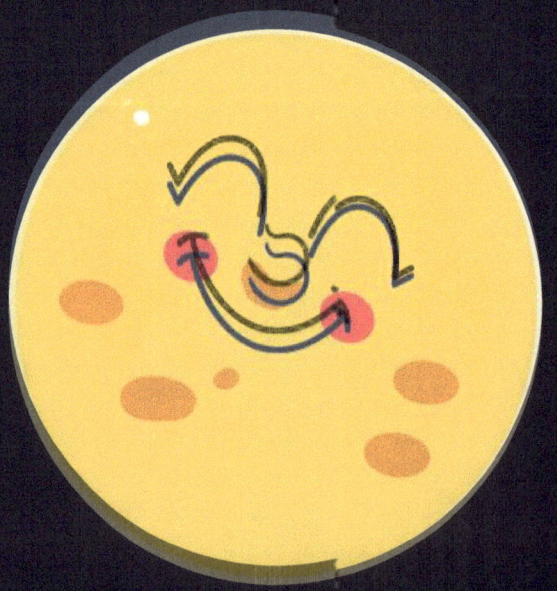

With hearts full of joy and minds brimming with curiosity, they looked forward to many more adventures and discoveries in the enchanted forest.

And as the sun set behind the trees, casting a warm glow over the forest, Maya, Oliver, and Sophie knew that their eclipse adventure was just the beginning of many magical journeys to come.

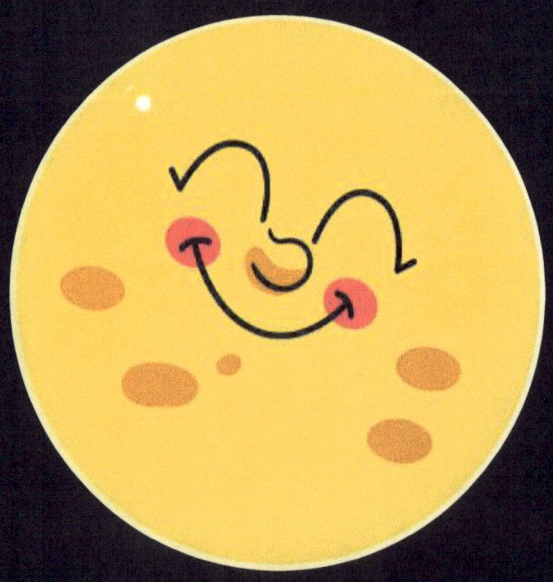

As the sun dipped below the horizon, signaling the end of another day in the enchanted forest, Maya, Oliver, and Sophie gathered under the canopy of stars.

With a sense of fulfillment and unity, they reflected on their eclipse adventure and the bonds it had forged among them and their fellow creatures.

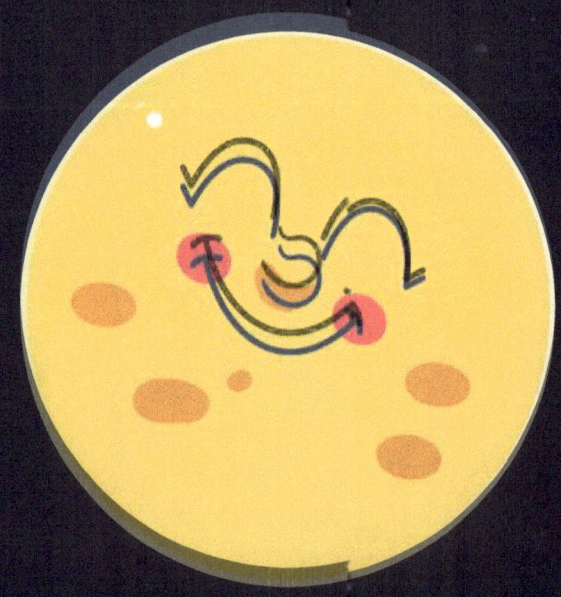

In that moment of tranquility, they realized that while the eclipse had passed, its magic would forever linger in their hearts, a reminder of the beauty and wonder that surrounded them.

With a promise to continue exploring, learning, and sharing their adventures with one another, Maya, Oliver, and Sophie bid each other goodnight, knowing that they had become not just friends, but guardians of the forest's mysteries.

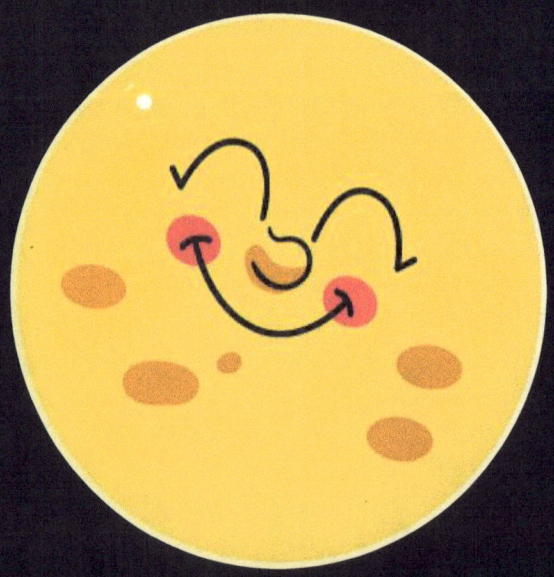

And as they drifted off to sleep, surrounded by the soft rustle of leaves and the gentle hoots of owls, they dreamed of the countless adventures that awaited them in the ever-enchanting world they called home.

THE END

www.ingramcontent.com/pod-product-compliance
Lightning Source LLC
Chambersburg PA
CBHW051936210526
45473CB00006B/2272